HOW TO BUILD
A PIPE BENDING MACHINE

Written and illustrated by
Vincent R. Gingery

Published by
David J. Gingery Publishing LLC
P.O. Box 318
Rogersville, MO 65742

Web: http://www.gingerybooks.com

Email: gingery@gingerybooks.com

Printed in the USA

FIRST EDITION
FIRST PRINTING 1998

Library Of Congress
Catalog Card Number 98-93412

ISBN 1-878087-21-5

NOTICE OF DISCLAIMER: There are serious hazards in the processes and procedures in this book. No attempt has been made to point out all of the dangers or even a majority of them. The methods, materials and procedures that are suggested in this manual were developed by a non-professional. The author is not an engineer or scientist and no claim is made to the propriety of the methods suggested in these pages. The reader is fully responsible for devising safe procedures for every operation.

Both the author and David J. Gingery Publishing hereby disclaim any liability for injury to persons or property that may result while using this book. Neither intends by this publication to explain all dangers known or unknown that may exist in the building and operation of the project

TABLE OF CONTENTS

Introduction

Bending a piece of pipe is often easy enough. The real challenge, is bending it to a nice even radius without kinking or flattening it at the bend. An experienced man can do the job by hand and even make it look easy. He fills the pipe with sand and clamps it in a vise. He knows exactly where to heat and cool the bend area, and at just the right place to make the bend. It may seem simple, but doing the job accurately by hand requires careful work and skill. It helps a great deal to have a machine that will do the job for us. And we're in luck, because in this project we will be building just such a machine.

As you can see in the photo below, it's a simple machine that's simple to build. And it mounts easily at the end of the work bench. By the use of leverage and a couple of dies shaped to fit the material, the bending job becomes quick and easy.

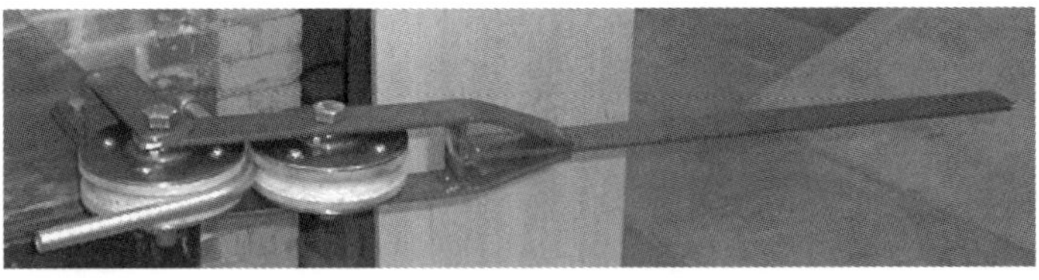

Figure 1

The design & how it works

I have needed a pipe bender from time to time in the shop, but really couldn't justify the cost of a commercial machine. I had thought about building my own machine for quite some time, but the perceived problems with making my own dies kept me from it. I was certain that they would have to be made either from castings or from thick chunks of steel machined on the lathe. It was when I considered hardwood that the idea became plausible. I was skeptical at first, but decided to give it a try. I made my first set of dies out of red oak sandwiched between 10 gauge sheet steel and shaped them to bend 1/2" conduit. I am happy to report that they worked excellent.

The overall design of the machine is patterned somewhat after a drawing I saw in an old book by Hiscox titled, "Mechanical Appliances, Mechanical Movements and Novelties of Construction." The drawing in figure 2 shows the various parts of the machine. In operation, the pipe to be bent, rests between the two dies, those being the forming die and the bending die. The pipe extends out through the holding bracket and is held in position by a clamp which securely holds the fixed end of the pipe during the bending operation.

The machine is operated by pulling the bending lever in the direction shown in the drawing. This puts the bending die in motion, causing it to roll around the forming die, bending the pipe as it goes.

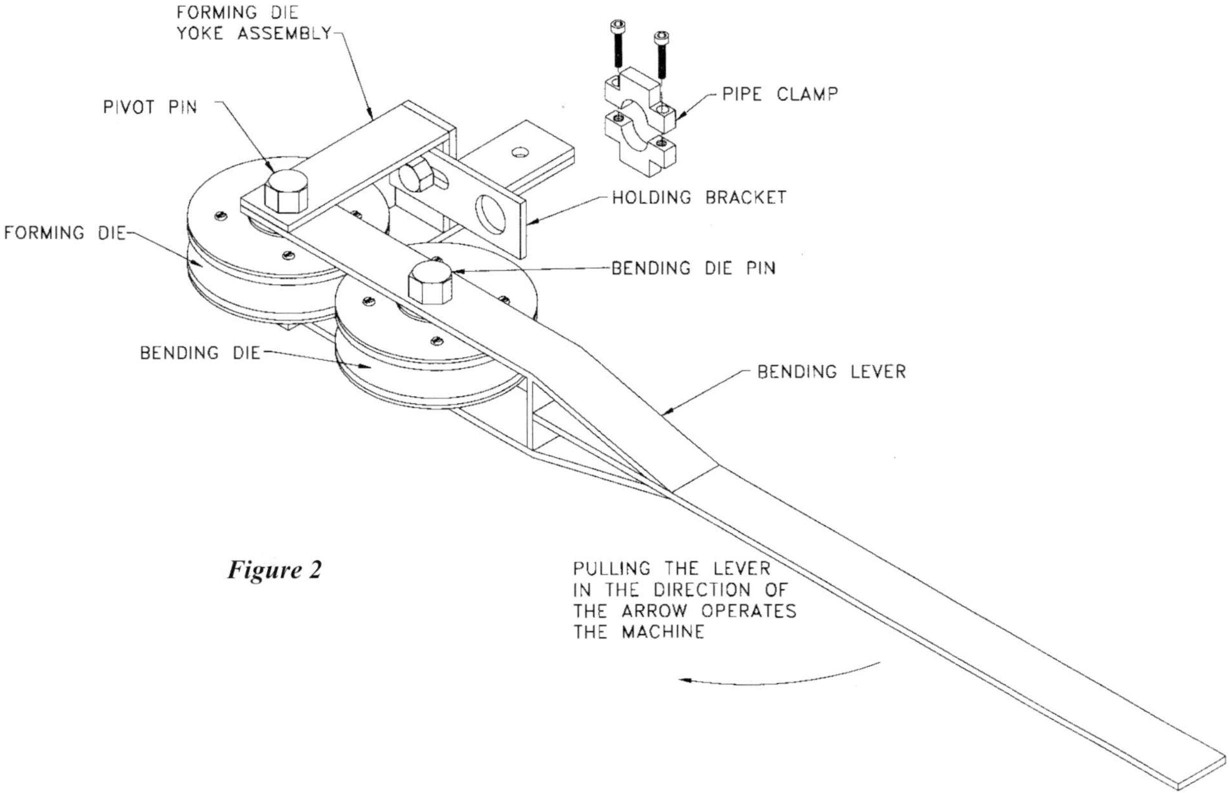
FORMING DIE
YOKE ASSEMBLY
PIVOT PIN
PIPE CLAMP
HOLDING BRACKET
FORMING DIE
BENDING DIE PIN
BENDING DIE
BENDING LEVER
PULLING THE LEVER
IN THE DIRECTION OF
THE ARROW OPERATES
THE MACHINE

Figure 2

The bending die does the work while the forming die is sized to give the proper radius to the pipe being bent. Both dies are grooved to fit the outside radius of the pipe being bent and they are designed to work with a minimum amount of clearance between them. That clearance being no more than 1/16". The close fit of the dies is important because the closer they fit together, the less likely the pipe will flatten or kink as you are bending it.

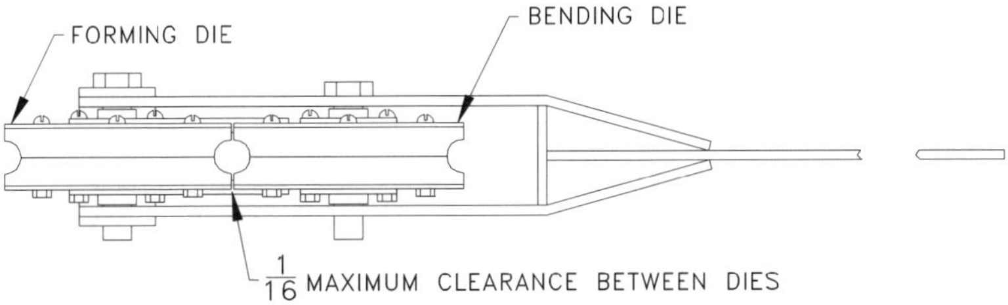

Figure 3 *Side view of machine.*

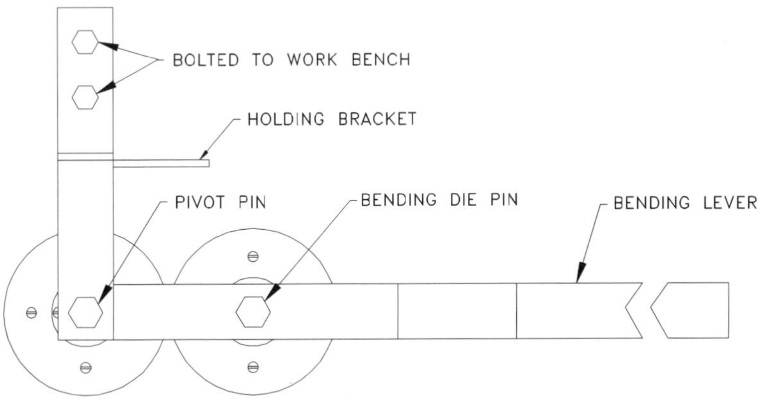

Figure 4 *Top view of machine.*

Tools & material

The frame of the machine is built from 1/4" x 2" hot roll steel flat bar. A bandsaw or hacksaw will be needed to cut the material to length. And a welder capable of 75 amps or more will be needed to weld the frame together.

There are several holes in the frame with diameters of up to 1". And one hole with a diameter of 1-1/4”. If you have a set of drill bits that include the larger sizes you can step drill the holes. Step drilling means to start the hole with a small drill, say 1/8" then gradually increase the hole size with progressively larger drills. Another method (the one I used) may be to drill the larger holes using a good quality metal cutting hole saw. Holes drilled with hole saws are not as accurate as those drilled with a drill bit. So if you use a holesaw, just be aware that your 3/4” hole may end up slightly oversized.

The bending and forming dies are made from hardwood sandwiched between 2 pieces of 10 gauge sheet steel and held together with 1/4-20 round head utility bolts. The hardwood portion of the die is 1-1/2" thick. I used red oak for my dies and gained the proper thickness by gluing two 3/4" thick boards together. The 10 gauge sheet steel will be too heavy to cut with tin snips so you will need a jig saw with a metal cutting blade or a bandsaw to cut the circular shape.

Both dies have a hub mounted in their center. The hub is made from 1" diameter cold roll steel round rod drilled or bored to a 3/4" inside diameter. A 1" flat washer is brazed to the end of each hub and serves as a mounting flange so you will need a torch capable of brazing. The pivot pin that the bending lever hinges on is made from the un-threaded shank portion of a 6" long, 3/4" machine bolt.

Both dies are grooved to fit the pipe that is being bent. This requires a couple of wood turning tools. Those being a parting tool and a 1/2" round nose chisel. You will also need a method of turning the form such as a wood or metal lathe. If you don't have a lathe, other methods could be used such as a drill press or an electric motor with an improvised tool rest. We'll discuss that more later.

The machine was designed around material that I already had in my shop. Don't be afraid to substitute if necessary. Certainly heavier is better in a project like this. For instance, the frame could be built from 1/4" x 3" flat bar. Or 3/8" x 2", 1/2" x 2" etc..

All of the hardware items you will need to build the machine can be found at the hardware store. I purchased the hardwood for the dies at the lumber yard. Chances are, the 1/4" x 2" H.R.S. flat bar for the frame can be purchased at the hardware store too. But, if you have a steel supply company in your area, check with them first because it will probably be cheaper. Ask about drops. Drops are pieces of material left over from jobs and companies often sell these pieces at reduced prices. Avoid cut off fees. In most cases they will charge you an arm and a leg to cut the material to length. If you find a piece of material longer than what you need, take it home and cut it to length yourself.

For those of you who don't know, Cold roll steel (C.R.S.) and hot roll steel (H.R.S.) refer to a particular grade of steel and how it was formed. Hot roll steel is a general utility grade steel. It has a much rougher surface and is cheaper to buy than C.R.S.. In our case, we will be using H.R.S. to build the frame and C.R.S. for the hub.

Material list

1 piece, 1/4" x 2" H.R.S. flat bar 12" long. (Lower yoke arm, figure 5)

1 piece, 1/4" x 2" H.R.S. flat bar 3-1/8" long. (Yoke spreader, figure 7)

1 piece, 1/4" x 2" H.R.S. flat bar 6-3/4" long. (Upper yoke arm, figure 6)

1 piece, 1/4" x 2" H.R.S. flat bar 7-1/2" long. (yoke brace, figure 8)

1 piece, 1/2" x 2" H.R.S. flat bar 7" long. (Pipe holding bracket, figure 9)

2 pieces, 1/4" x 2" H.R.S. flat bar 17" long. (Top & bottom bending lever arms, figure 19)

1 piece, 1/4" x 2" H.R.S. flat bar 27" long. (Bending lever handle, figure 21)

1 piece, 1/4" x 2" H.R.S. flat bar 2-5/8" long. (Bending lever spreader, figure 20)

2 pieces, 1" diameter C.R.S. round rod 2-5/8" long. (Hub, figure 25)

1 piece, 1" diameter x 4" long C.R.S. round rod. (Work arbor, figure 31)

Five, 1" flat washers. (Hub flanges, figure 26)

2 pieces, 6" diameter hardwood 1-1/2" thick. (Bending dies to bend 1/2" pipe or conduit, figure 33)

4 pieces, 6" diameter 10 gauge sheet steel. (Outer facing for dies, figure 34)

Four, 1/4-20 x 2-1/4" utility bolts with nuts. (To bolt hub flange to die)

Eight, 1/4-20 x 2" utility bolts with nuts. (Used to hold the die together)

Two, 3/4" x 6" machine bolts. (Pivot pins, figure 49)

One, 1/2-13 x 3/4 machine bolt with flat washer. For attaching the holding bracket. (figure 17)

Chapter 1

The forming die yoke assembly

Gather up the following material for the forming die yoke assembly.

A 12" length of 1/4" x 2" H.R.S. flat bar for the lower yoke arm shown in figure 5. Locate and drill the two 1/2" holes as shown in the figure.

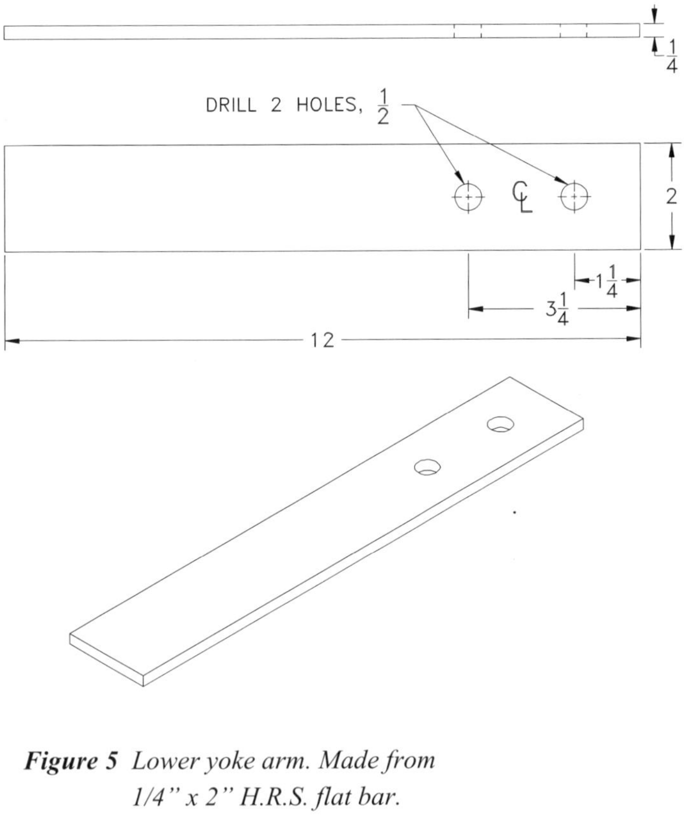

Figure 5 *Lower yoke arm. Made from 1/4" x 2" H.R.S. flat bar.*

A 6-3/4" length of 1/4" x 2" H.R.S flat bar for the upper half of the yoke arm as shown in figure 6.

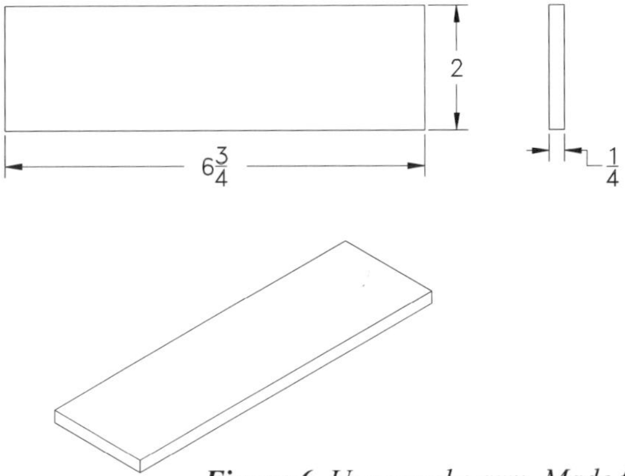

Figure 6 *Upper yoke arm. Made from 1/4" x 2" H.R.S. flat bar.*

A 3-1/8" length of 1/4" x 2" H.R.S. flat bar for the yoke spreader shown in figure 7. Locate drill & tap the the center hole 1/2-13.

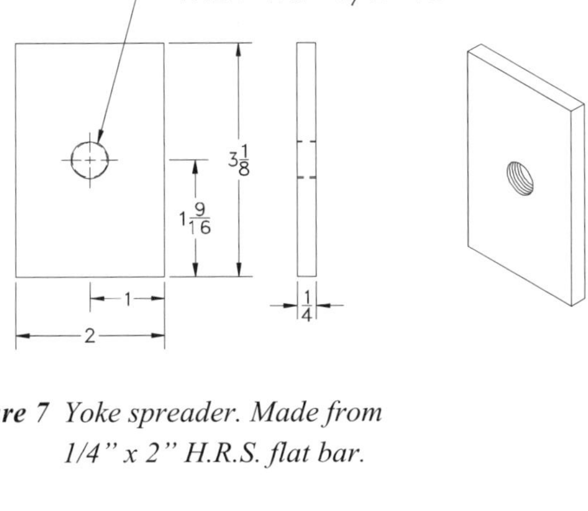

Figure 7 *Yoke spreader. Made from 1/4" x 2" H.R.S. flat bar.*

A piece of 1/4" x 2" H.R.S. flat bar 7-1/2" long for the yoke brace shown in figure 8. Scribe the bend line on the brace as shown in the figure. You can bend the brace by clamping it in a vise. Align the bend line with the top of the vise jaws. Use a large adjustable wrench to grip the end and make the bend. See figure 8A. You can see how the brace is attached to the yoke in figure 15.

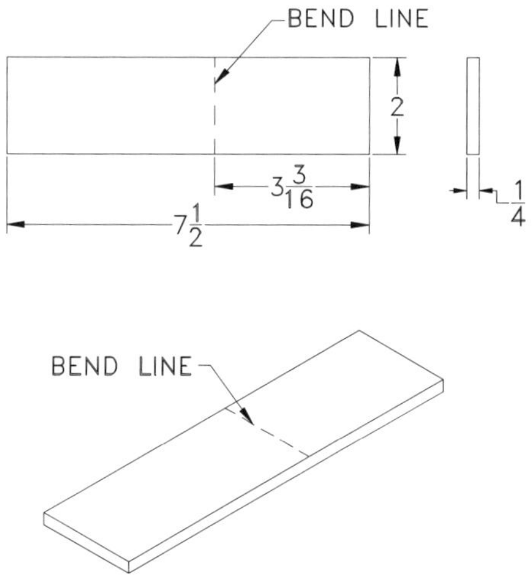

Figure 8 *Yoke brace. Made from 1/4" x 2" H.R.S. flat bar.*

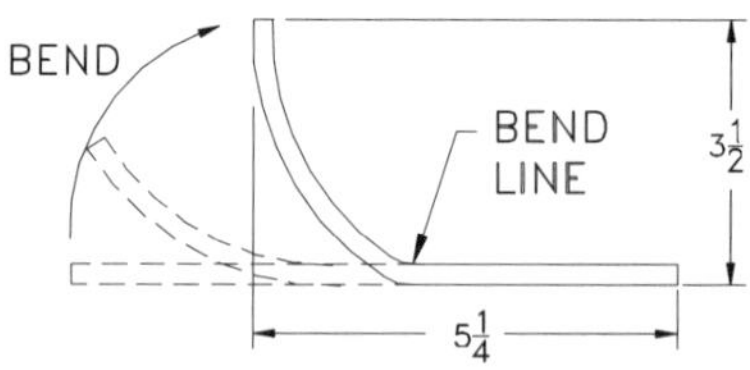

Figure 8A Clamp the brace in a vise and bend as shown above.

Next you will need a 7" length of 1/2" x 2" H.R.S. flat bar for the pipe holding bracket shown in figure 9. Drill a 1/2" diameter hole, then cut the 1/2" wide slot in the bracket as shown. Drill the 1-1/4" hole in the pipe bracket. Large holes are difficult to drill with a normal drill bit and besides, a set of drill bits of this size are expensive. A good quality metal cutting hole saw seemed to work best for this hole. See figure 10.

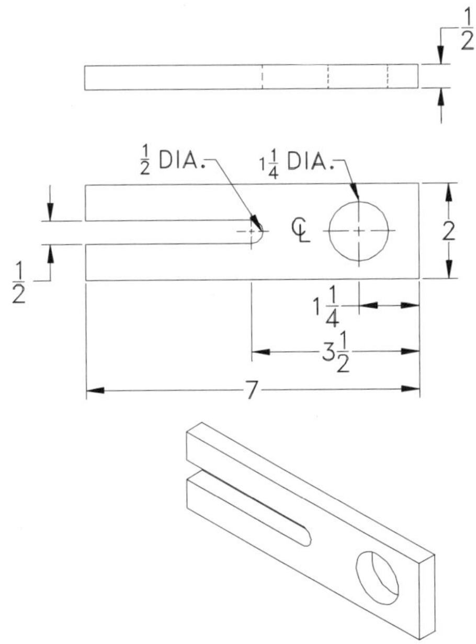

Figure 9 *Holding bracket. Made from 1/2" x 2" H.R.S. flat bar*

Figure 10

Using a 1-1/4" hole saw to drill the hole in the holding bracket.

When the material is gathered up you can put the yoke assembly together.

Begin by scribing a line on the lower yoke arm, 6-1/2" from the end as shown in figure 11.

Position the spreader on the lower yoke arm, aligning the front edge of the spreader with the scribed line and weld in place. See figure 12.

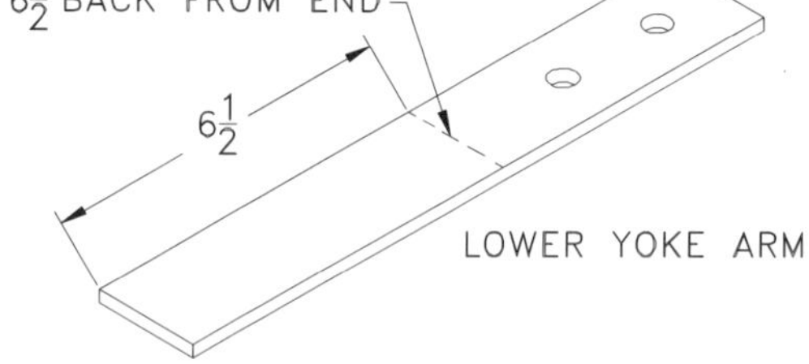

Figure 11 *Step 1*

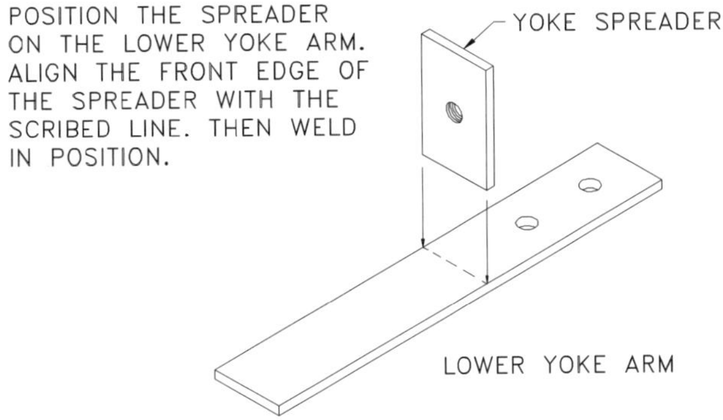

Figure 12 *Step 2*

Position the upper yoke arm on the top edge of the spreader. Align back edge of the yoke arm with the back edge of the spreader. Weld in position. See figure 13.

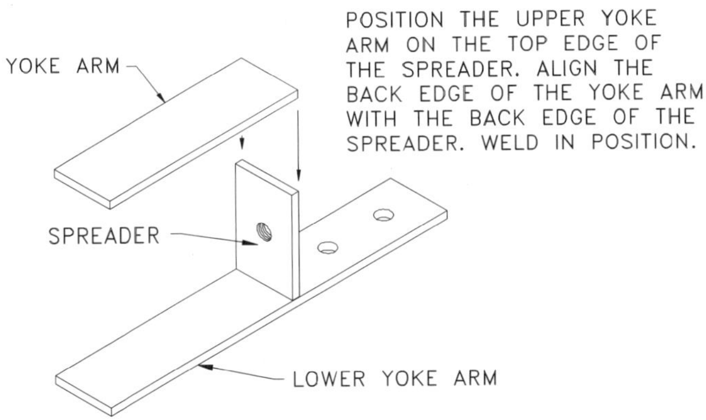

Figure 13 *Step 3*

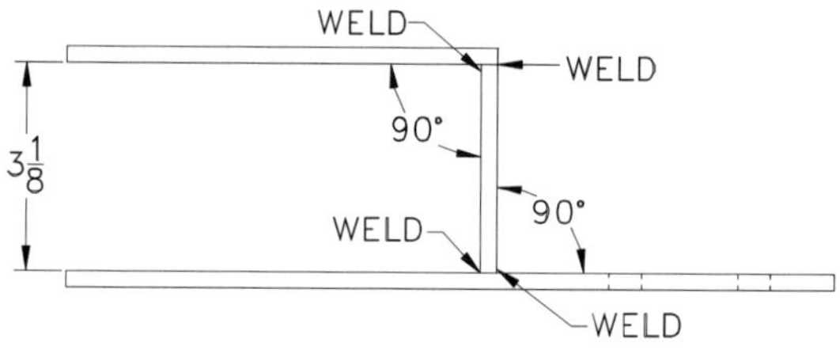

Figure 14 *Side view showing weld locations.*

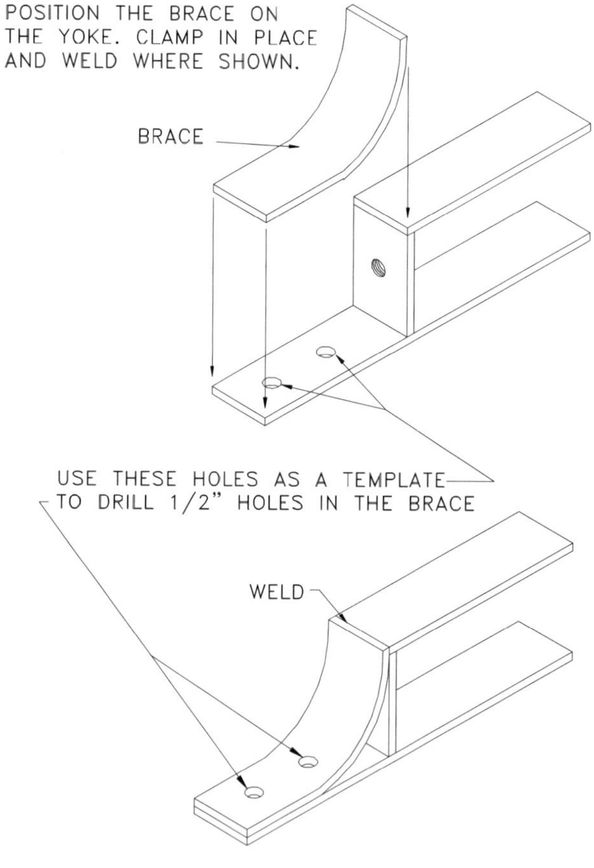

Figure 15 *Step 4*

Position the yoke brace from figure 8A on the yoke. Clamp it in place and weld where shown. Use the two 1/2" holes in the lower yoke arm as templates to drill 1/2" holes through the brace.

Locate the position for the 3/4" holes, 1" back from the end of the yoke. See figure 16. Step drill these holes to 3/4". The holes need to be aligned, so it is best to drill them through at the same time. The top hole becomes a template for the bottom hole. Put a spacer between the yoke arms while drilling to keep the top bar from bending down under the pressure of the drill bit.

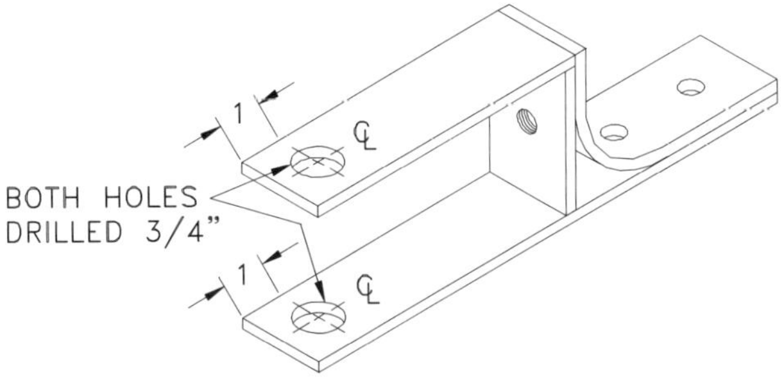

Figure 16 *Step 5*

Fasten the holding bracket to the spreader with a 1/2 -13 x 3/4" machine bolt and flat washer as shown in figure 17.

See figure 18 for a photo of the finished yoke assembly.

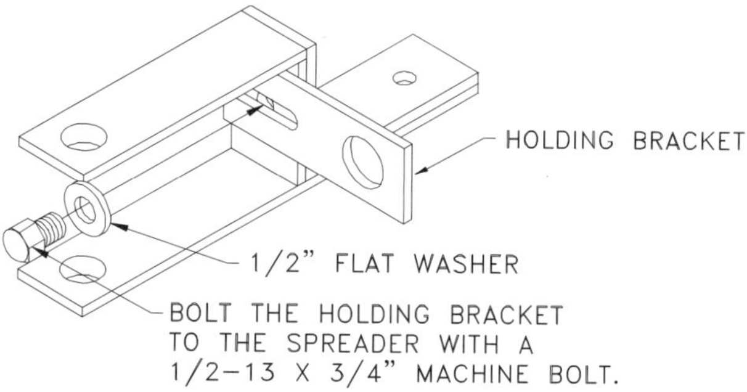

Figure 17 *Step 6*

Figure 18 *Completed yoke assembly*

Chapter 2

The bending lever

Gather up the following material for the bending lever.

Two 17" long pieces of 1/4" x 2" H.R.S. flat bar for the top and bottom lever arms shown in figure 19.

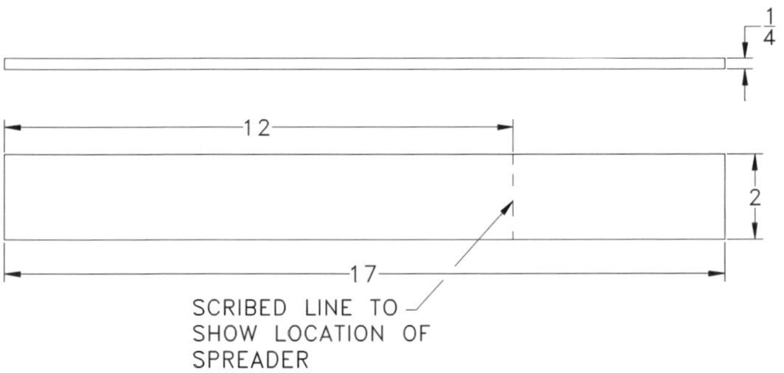

Figure 19 *Top and bottom arms for the bending lever. Make 2 pieces from 1/4" x 2" H.R.S. flat bar.*

A 2-5/8" length of 1/4" x 2" H.R.S. flat bar for the spreader shown in figure 20.

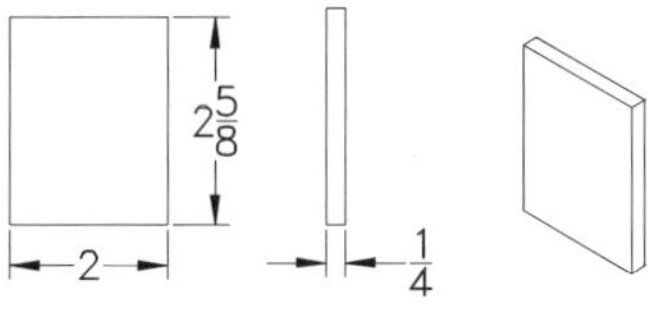

Figure 20 *Spreader for the bending lever. Make it from 1/4" x 2" H.R.S. flat bar.*

A 27" length of 1/4" x 2" H.R.S. flat bar for the handle shown in figure 21.

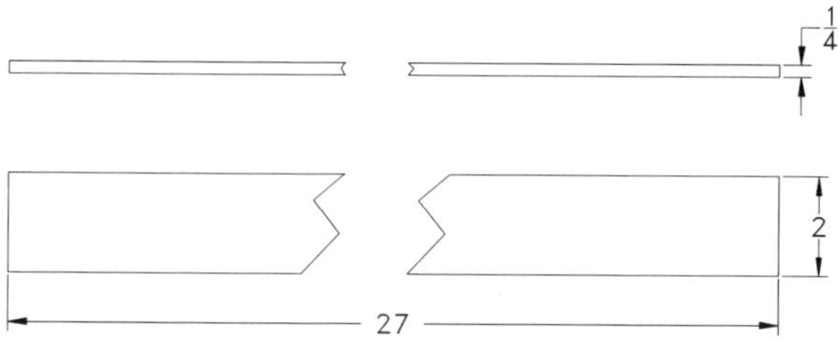

Figure 21 *Bending lever handle.*
Make it from 1/4" x 2" H.R.S. flat bar.

Assemble the lever. The exploded view in figure 22 shows the assembly. Align the front edge of the spreader with the scribed lines on the top and bottom bending lever arms. Clamp in position and weld where shown in figure 23.

Bend the back ends of the lever arms down to the approximate center of the spreader. Figure 23 shows it best. You can accomplish this by clamping the assembly in a vise and using a large crescent wrench adjusted to fit the end of the arm to do the bending.

Next position the lever handle centered against the back side of the spreader. Weld the handle to the spreader and the lever arms to the handle. Locate and drill the two 3/4" holes in the end of the lever arms the same way you did those in the yoke assembly. See figure 24.

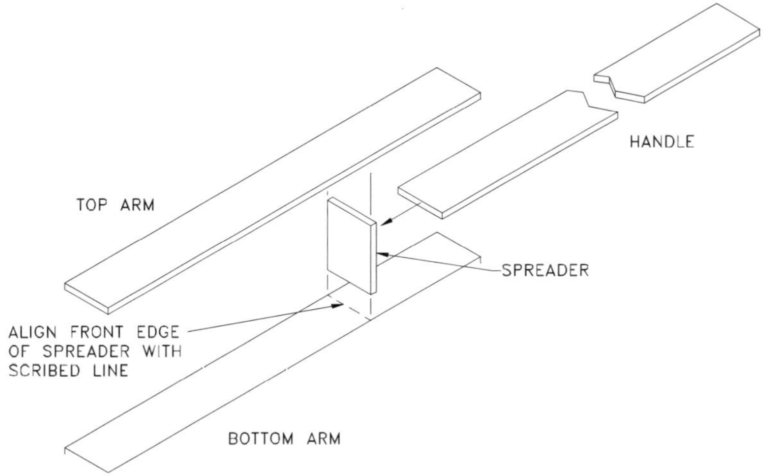

Figure 22 *Bending lever assembly*

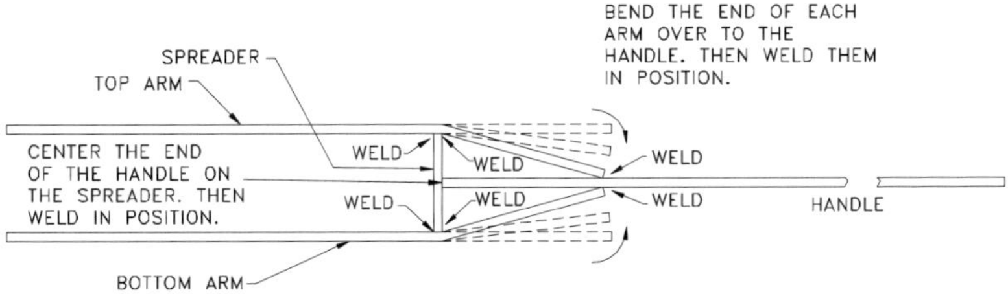

Figure 23

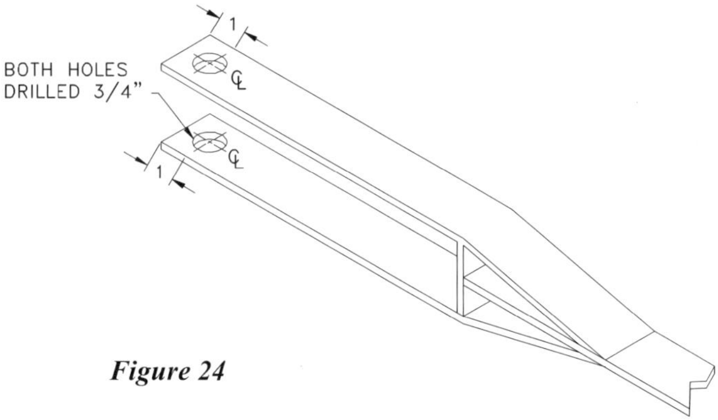

Figure 24

Chapter 3

The hubs

You will need 2 hubs for the machine. One for each die. The hubs are interchangeable between the dies, so dies of different sizes will be able to use the same hubs.

Gather up the following material to make the hubs.

Two, 2-5/8" long pieces of 1" diameter C.R.S. round rod for the hub blanks shown in figure 25.

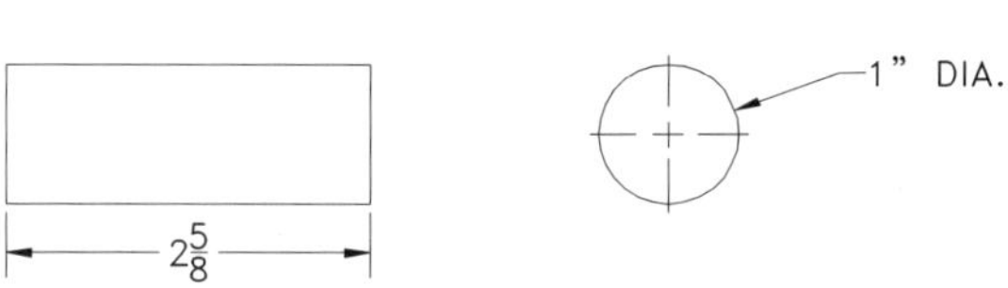

Figure 25 *Hub. Make 2 from 1" diameter C.R.S. round rod.*

Five, 1" flat washers. These will be the hub flanges. The 1/4" holes in all 5 must match, so locate and drill the two 1/4" holes in one and then use it as a template to locate and drill the holes in the other 4. See figure 26.

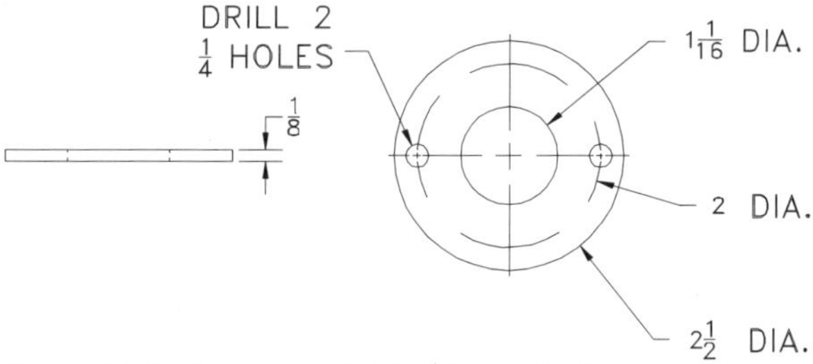

Figure 26 *Hub flanges. Make 5 matched flanges from 1" flat washers.*

Scribe a line around each hub, 3/8" from one end. Align the back edge of one of the flat washers with the scribed line. Be sure the hub is perpendicular to the washer. Then braze the washer to the hub from the top side only. Repeat the process for the other hub.

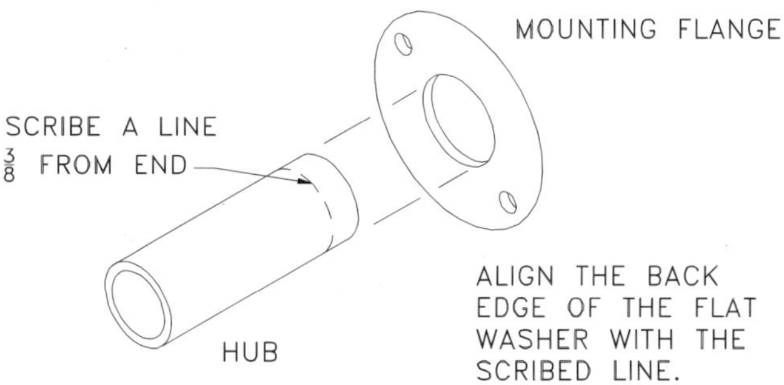

Figure 27

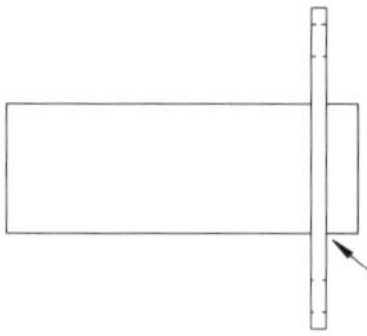

Figure 29

Bore or drill a 3/4" hole through the center of each hub. If you have a lathe it is a simple matter to chuck a hub in the 3-jaw and bore the hole. If you do not have a lathe you may be able to use a drill press. If you use a drill press, begin by clamping the hub securely in a drill vise. The holes must be drilled straight, so make sure the hub is aligned with the drill spindle, then carefully step drill the holes. When drilling the large holes, it is best to set the drill at its slowest speed preferably less than 200 RPM.

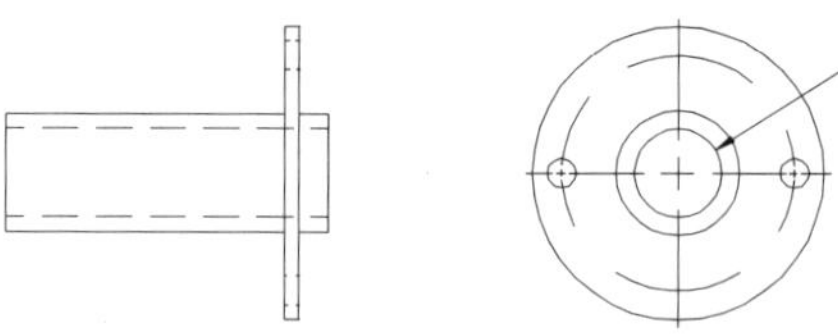

Figure 30 *Finish hub by drilling or boring 3/4" through the center*

The work arbor

The work arbor is made just like one of the hubs except the length of the shaft will be 4". Also, the work arbor does not have a 3/4" hole bored through it. To assemble the arbor, simply braze a flange to the end of the shaft. See figure 31A for a photo.

The arbor shown here is made to fit in the lathe chuck. If you plan on using a different method of turning the die form, you will want to make the arbor somewhat different to fit your needs. For example, look ahead to page 34 which shows how to use a drill press as a wood lathe.

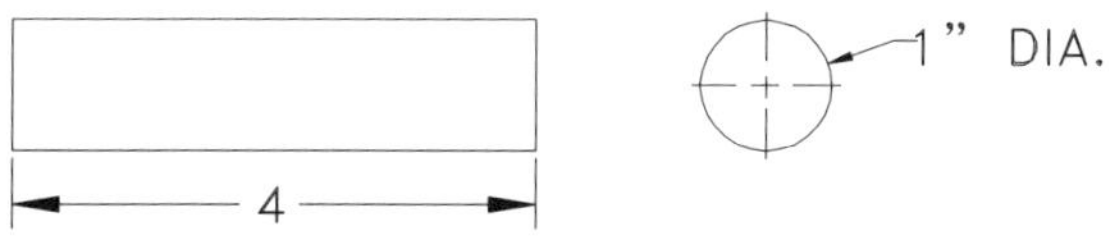

Figure 31 *Shaft for the work arbor. Make from 1" diam. x 4" long C.R.S round rod.*

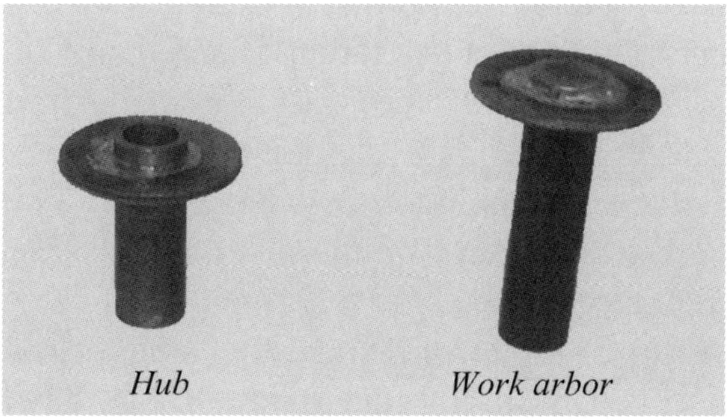

Figure 31 A

Chapter 4

Designing the dies

Figure 32 *A set of dies for bending 1/2" electrical conduit*

The dies in this project are sized to bend 1/2" electrical conduit. There are many different types and sizes of pipe, so many die sizes and combinations are possible. Die construction will be the same no matter what size pipe, square tubing, solid round rod or flat bar that you are bending so you can use the following example as a guideline.

There are some important things to consider before making your dies. First of all, there is a general rule of thumb that says thin walled pipe should not be bent to a radius of less than 4 times its diameter. That's because at a smaller radius the wall of the pipe has an increased risk of flattening or kinking as it is being bent. This is just a general rule of thumb though. Thicker walled pipe can be bent to a smaller radius. And thin walled pipe can be bent to a smaller radius too, if you fill it with dry sand and cap the ends before you bend it. Dry sand helps support the walls of the pipe preventing them from collapsing. In our example, the outside diameter of the 1/2" conduit we will be bending measures just a little less than 3/4". So in this case, the radius of our forming die will be 3". (3/4" x 4" = 3") If you were to bend 1" pipe the radius of the forming die should be no less than 4".

The minimum diameter requirements above do not apply to the bending die. For larger diameter pipe I think the bending die should have at least a 3" radius. And in our example the radius of the bending die will be 3". But let's say you were going to bend a 1" diameter piece of pipe. The radius of the forming die could be 4" and the radius of the bending die could be 3". You will find that there can be several different combinations. Also remember that in designing your dies, the bending die must be sized for minimum clearance between it and the forming die. And, keep in mind when sizing your dies that the bushing in the bending die has to align with a pivot pin hole in the bending lever.

Making the dies

You will need to make 2 dies. Each is made from a 6" diameter x 1-1/2" thick piece of hardwood. See figure 33. I used red oak, but other hardwoods such as maple, white oak etc. should work as well. The oak boards I found at the lumber yard were only 3/4" thick so to gain the 1-1/2" thickness required, I glued 2 boards together with wood glue.

Scribe a 6" diameter circle in the board for each die and cut the forms out with a jig saw. Then drill a 1" hole through the center of each die as shown in the figure.

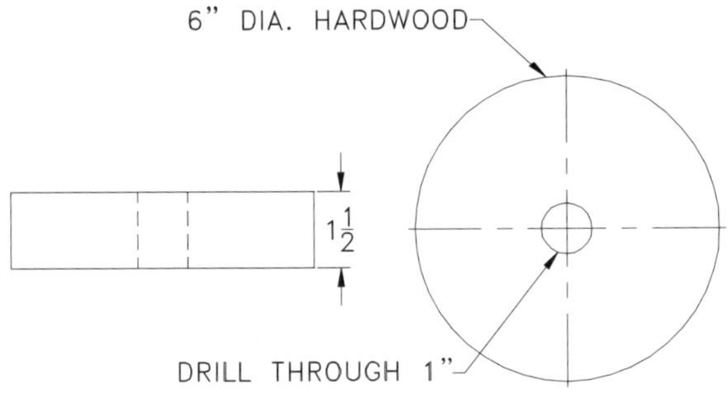

Figure 33 *Die blank. Make 2 from 1-1/2" thick ,6" diameter hardwood.*

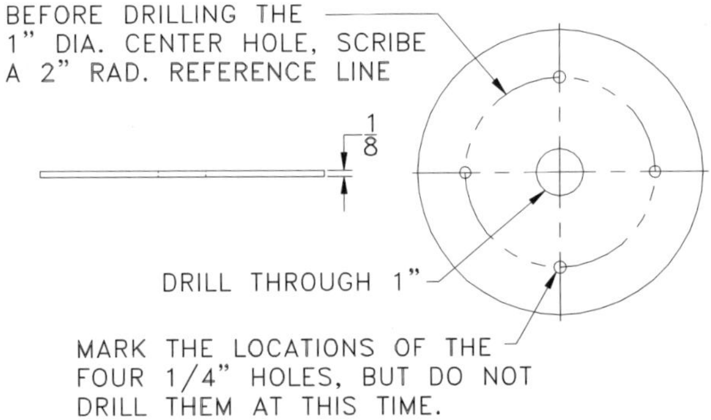

Figure 34 *End plates for the dies. Make 4 from 6" diameter, 10 gauge mild steel.*

To add strength to the dies they are sandwiched between two 6" diameter 10 gauge steel discs referred to as end pieces. See figure 34. You will need 4 of them. You can cut the 6" diameter forms from a sheet of 10 gauge mild steel using a heavy duty jig saw with metal cutting blade. A band saw will work too, or a cutting torch.

Scribe the 2" radius circle on each end piece and layout the location of the four, 1/4" holes that will be drilled later. Then drill a 1" hole carefully centered in each piece.

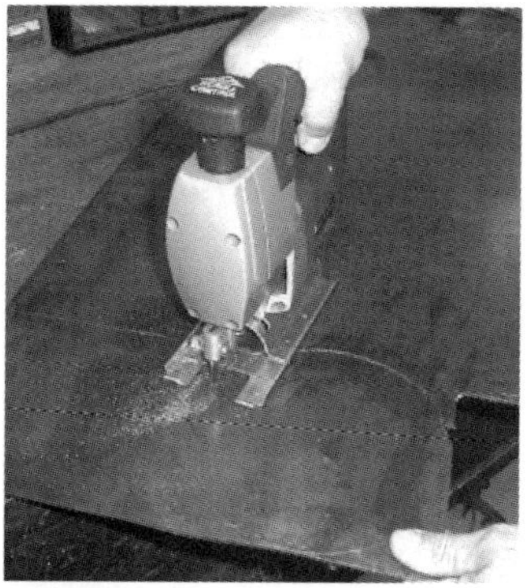

Figure 35 *Cutting an end piece using a jig saw.*

Figure 36 *Shown here are the wooden die, the two 10 gauge end pieces and the work arbor ready for assembly*

Sandwich the wooden die between the two 10 gauge end pieces and insert the work arbor through the 1" center holes of each piece. See figure 37. Using the 1/4" holes in the flange of the work arbor as a template, drill 1/4" holes through the complete die assembly. Then bolt the work arbor to the die with 1/4-20 x 2-1/4" round head utility bolts. See figure 38.

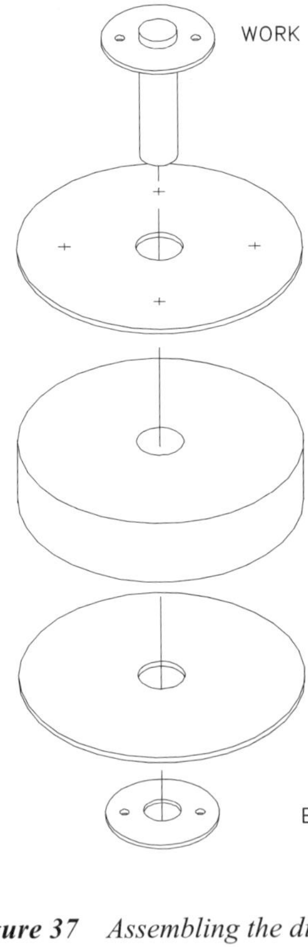

Figure 37 *Assembling the die.*

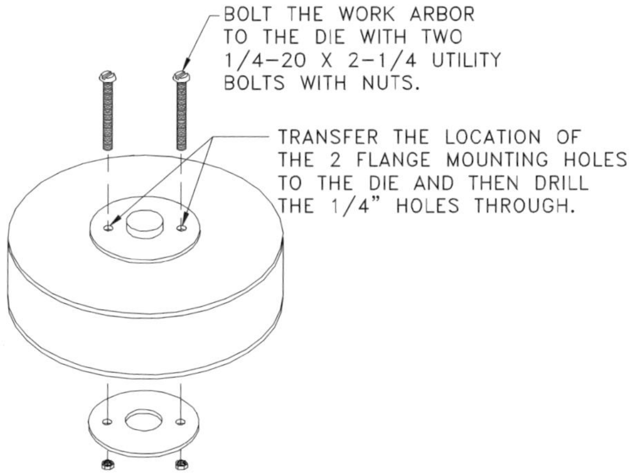

Figure 38 *Bolting the work arbor to the die*

Drill the four 1/4" holes marked at the 2" radius. Finish bolting the die together with four, 1/4-20 x 2" round head utility bolts. See figure 39.

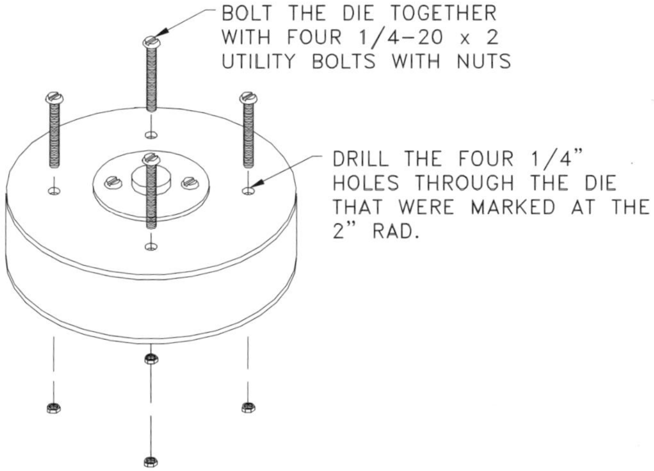

Figure 39 *Bolting the die together.*

When you are through assembling the dies you are ready to turn the groove in each one to fit the pipe you are going to bend. To do this you will need a lathe parting tool, a round nose tool and a method of turning the work. Since I have a metal lathe that is what I used. You will notice in the photos that I used my 1/2" boring bar as a tool rest. If you have a wood lathe you're in good shape and probably won't need the work arbor. It will just be a matter of centering the work on the face plate and bolting the die in position.

If you do not have a lathe of any kind, you could use a drill press fitted with a make shift tool rest. Or maybe an electric motor mounted on the work bench and connected to a counter shaft by belt and pulley to reduce the speed of the motor. The work could be mounted on the end of the counter shaft for the turning operation.

Using a drill press as a wood lathe

I didn't try the electric motor idea, but I did give the drill press idea a try. It was kind of awkward because the die was in a horizantal position and I was working from the side, but it did work. I learned that it is important to support both ends of the arbor. If you don't, the taper shank could come loose from the spindle and send your work flying across the room. Also, start the drill at its slowest speed until you become familiar with the operation. The drawing and photos on page 35 show the set up.

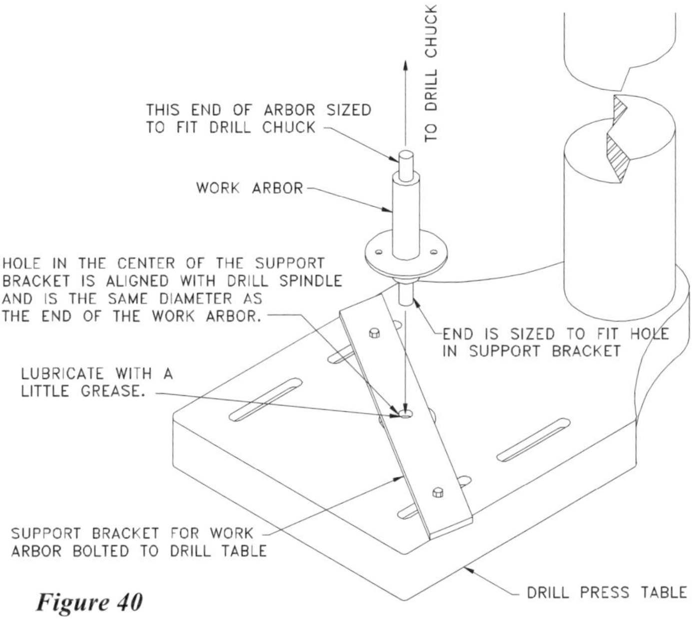

Figure 40

Figure 40a *A short piece of angle iron bolted to the drill table serves as the tool rest. In this photo, the die is not shown bolted to the arbor so that you can see the set up.*

Figure 40b

The following steps will show you how to turn the groove in each die. The cut we will be making into the side of each die is called a concave cut.

Step 1. Prepare the pattern. Our pattern will be a short piece of the pipe we will be bending. Cut the short piece of pipe in half length ways as shown in figure 41. This will be your pattern.

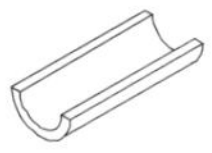

Figure 41
Pattern

Step 2. Draw a line along the exact center of the face of the die. The width of the cut will be the width of the pattern. (diameter of the pipe being bent) Lay out the width of the cut on the face of the die measuring equally from both sides of the center line. See figure 42.

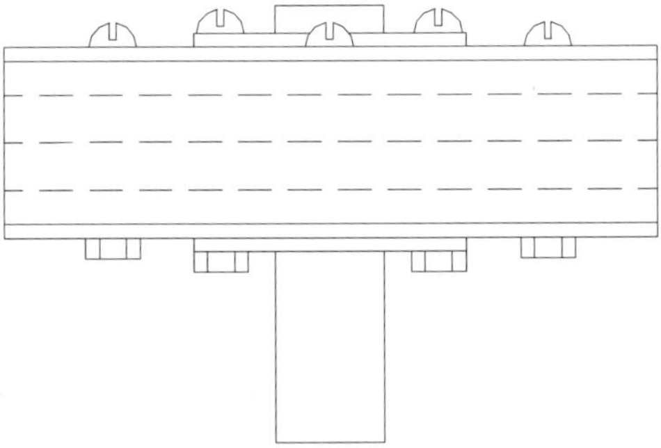

Figure 42 *Side view of die with work arbor installed. The center dotted line represents the center of the die and the outer dotted lines represent the width of the pipe to be bent..*

Step 3. The depth of cut will be about 1/32" less than the radius of the pipe being bent. Use the lathe parting tool to cut into the center of the wooden die until the proper depth is reached.

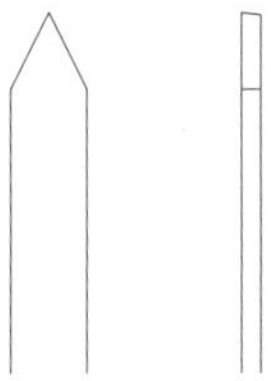

Figure 44
Lathe parting tool

Figure 45 *Using the parting tool*

Step 4. Use a 1/2" round nose chisel for making the concave cut. Hold the chisel level and perfectly flat on the tool rest and begin the cut a little inside the marked lines. Gradually work down to the bottom of the cut made with the parting tool.

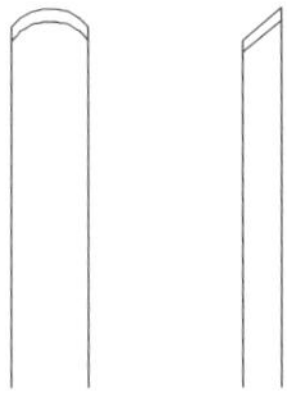

Figure 46
Round nose tool

Figure 47 *Using the round nose tool*

Figure 47A Checking the depth of the form with the pattern.

Test the form by using the pattern and when the proper depth and shape have been achieved, you are done.

Remove the work arbor from the die and replace it with a hub. Then mount the work arbor in the other die and repeat the above process. The work arbor can be used over and over again on other dies that you might happen to make.

CHAPTER 5
FINAL ASSEMBLY

Bolt the forming die yoke assembly to a work bench as shown. Use 1/2" bolts.

Figure 48 *The yoke bolted to the bench.*

Two 6" long 3/4" machine bolts are needed for the pivot pins. Cut off the threaded portion of each bolt.

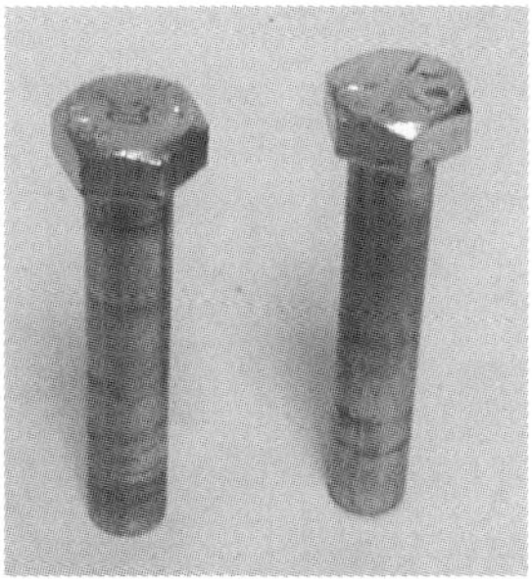

Figure 49 *Pivot pins*

Insert one of the dies between the arms of the bending lever. This die will be the forming die. Align the 3/4" hole in the hub of the forming die with the 3/4" holes at the end of the bending lever.

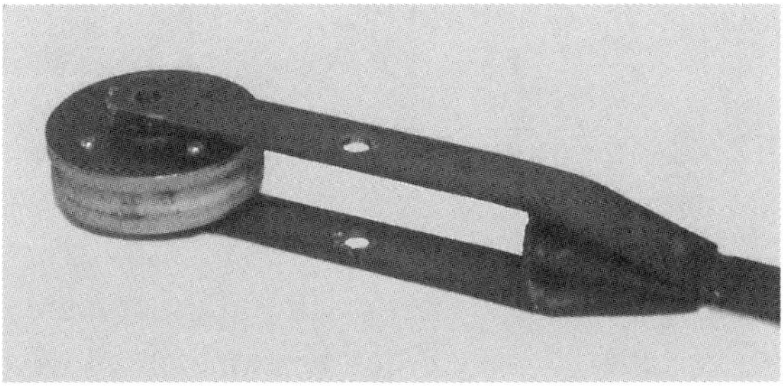

Figure 50 *Forming die in position*

Insert the bending lever with die in place between the yoke arms. Align the 3/4" hole at the end of the bending lever with the 3/4" holes at the end of the yoke arm. Insert a 3/4" pivot pin into the hole.

Figure 51 *Bending lever assembly with forming die positioned between yoke arms and ready for pivot pin.*

Figure 52 *Inserting the pivot pin*

Now we need to locate the position of the bending die. Center the bending die between the bending lever arms. Hold a short piece of the pipe you will be bending in the groove of one of the dies, then push the bending die flush against the forming die. There should be no more than 1/16" between the die forms. Determine the distance between the center of the forming die hub and the center of the bending die hub. Mark that location centered on the bending lever arm.

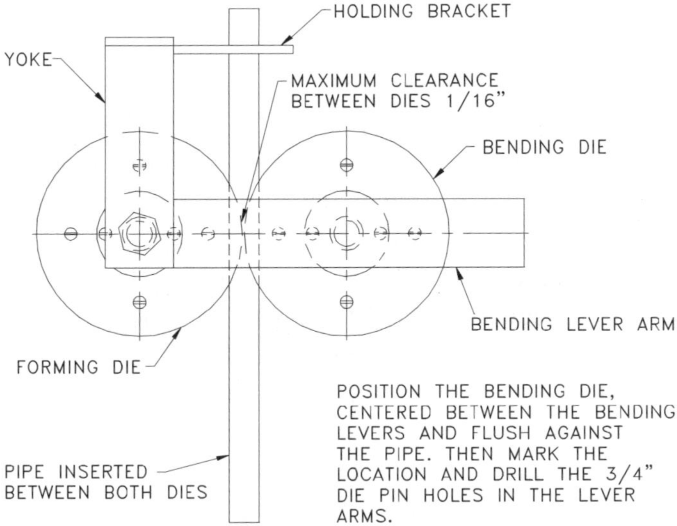

Figure 53 *Locating the position of the bending die.*

Remove the bending lever and drill the 3/4" hole for the bending die pin through both arms. The holes in both arms must be aligned with each other so drill through both at the same time. Use a spacer of some type to support the top arm while you are drilling.

When the bending die pin holes have been drilled, reassemble the machine as before. Position the bending die in place and insert the die pin.

Chapter 6

Operating the machine

To set up the machine, insert the pipe to be bent in the grooves between the dies and through the holding bracket. See figure 54.

Figure 54 *Bending die installed . The pipe we are going to bend is positioned between the dies.*

The machine will have a tendency to pull on the pipe rather than bend it. You will need to make some type of holding clamp or set screw collar to hold the pipe securely in place while you are making the bend. Very important! If the pipe slips during the bending operation it will kink at the bend. One idea and the one I used, (you may think of others) was a set screw collar sized to fit the pipe as shown in the drawing in figure 55 and in the photo in figure 57. The set screws will put small marks in the pipe. If the marks in the pipe pose a problem, you might consider making a clamp, something like the drawing in figure 56.

Figure 55 *Set screw collar to hold pipe.*

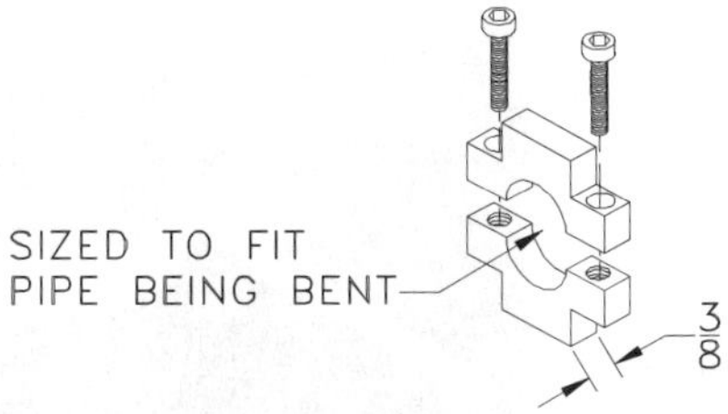

Figure 56 Clamping fixture to hold pipe.

When pipe is in position as in figure 54 and is held in place as in figure 57 you are ready to make the bend by pulling on the bending lever. Figure 58 shows the pipe in the process of being bent. Figure 59 shows the bending job complete and figure 60 shows the formed piece of pipe.

***Figure 57** Pipe being held in position by a set screw collar*

Figure 58 *the pipe in the process of being bent.*

Figure 59 *Bending job complete.*

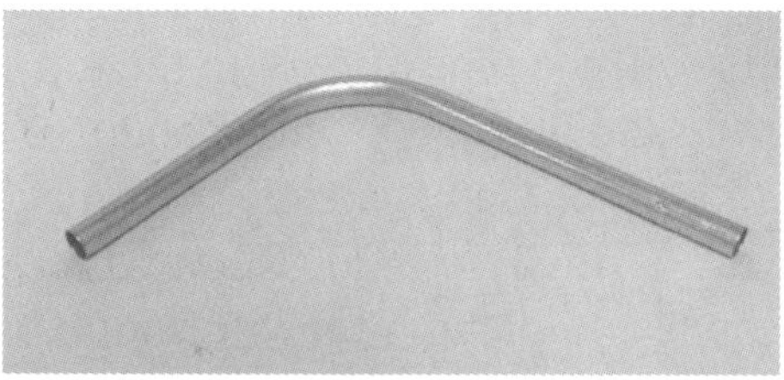

Figure 60 *Formed piece of pipe*

Conclusion

Congratulations on the completion of your new pipe bender. With proper dies it will do a very good job of forming a radius in pipe, solid round rod, flat bar and square tubing. The example in figure 61 shows what a set of dies for forming flat bar would look like.

In my experience, the machine formed near perfect bends in thick walled pipe such as water and gas pipe up to 1" in diameter. And it did a good job bending thin walled pipe up to 1/2" in diameter.

It was when I bent 3/4" electrical conduit (thin walled pipe) that I ran into problems. It had a tendency to kink and crush the pipe resulting in an unsatisfactory bend. I was able to remedy this problem and achieve excellent results by filling the pipe with dry sand before making the bend. (Must be dry sand). This is a common practice and it is done by capping off one end of the pipe and filling it with dry sand from the other end. One way of capping off the ends is with masking tape or duct tape. Be sure and pack the sand in tight by tapping the pipe against a solid surface. When you tap the pipe the sand will settle, so add more sand to fill the pipe even with the top edge. Repeat until sand no longer settles and it is level with the top edge. Cap the end, then make the bend as you normally would. Remove the sand after the job is complete. You will find that sand does a very good job of supporting the pipe wall and preventing it from collapsing during the bend. The result is a near perfect bend.

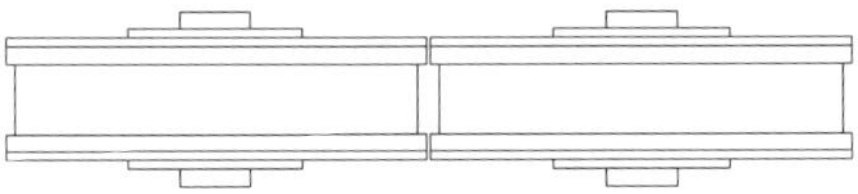

Figure 61 *A set of dies for forming flat bar.*

Formulas

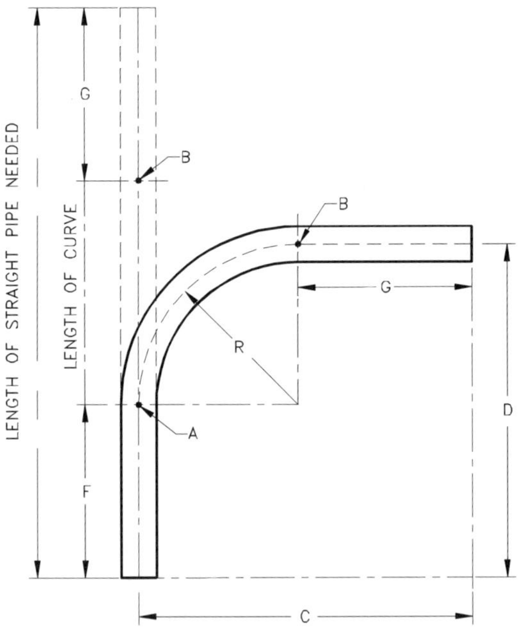

LENGTH OF PIPE NEEDED = F + 1.57 R + G

C = R + G

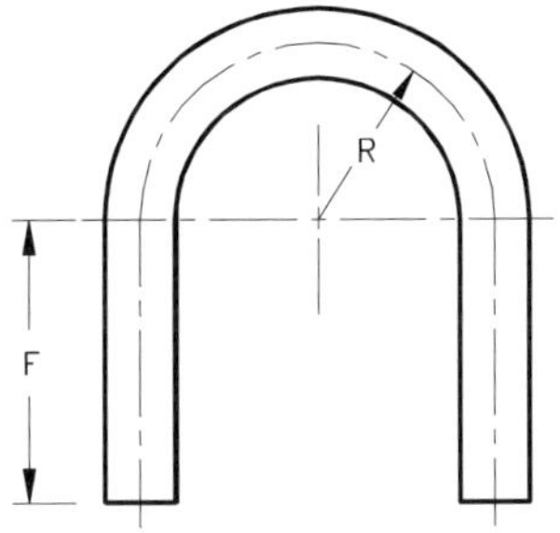

LENGTH OF PIPE NEEDED = 2F + 3.14R

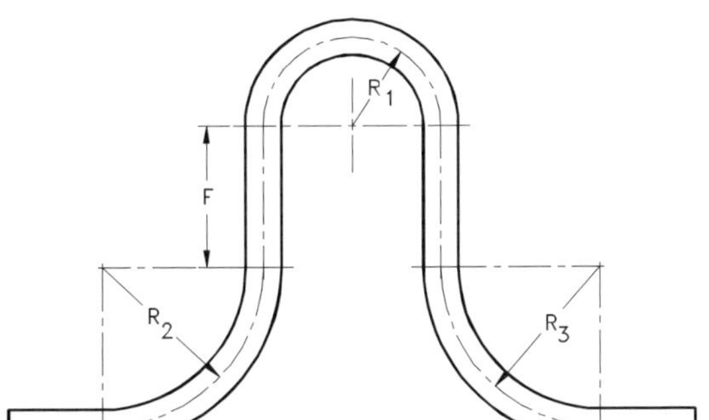

LENGTH OF PIPE NEEDED = 2F + 3.14 R_1 + G + 1.57 R_2 + H + 1.57 R_3

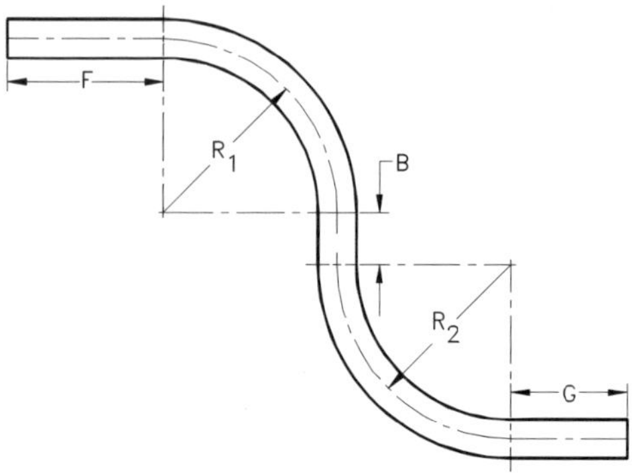

LENGTH OF PIPE NEEDED $= F + 1.57\ R_1 + B + 1.57\ R_2 + G$

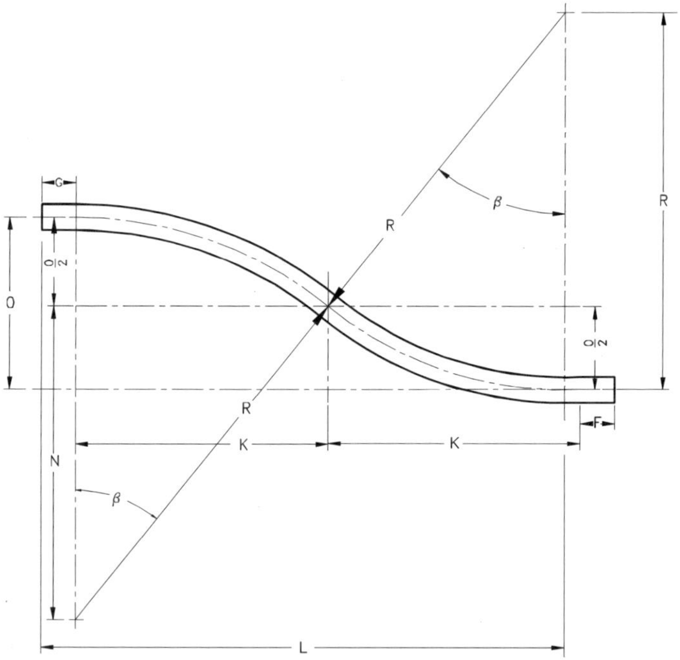

LENGTH OF PIPE NEEDED $= F + 2 \times 0.01745\ R\beta + G$

$N = R - \frac{O}{2}$

$\cos\beta = \frac{N}{R}$, FROM WHICH THE ANGLE β IS OBTAINED.

$K = N \times \tan\beta$

THE LENGTH OF EACH CIRCULAR ARC OF ANGLE $\beta = 0.1745\ R\beta$